Help your students engage in inquiry-based learning

National
Science
Education
Standards

Several sections in this book cite the applicable standards from the National Research Council's *National Science Education Standards.* Look for the initials **NSES** as you go through the book. To obtain a copy of *National Science Education Standards,* see **Professional Resources,** page 19.

General Activity Guidelines

Successful inquiry-based activities, such as those presented below and on the next page, should focus on these abilities:

- Identify questions that can be answered through scientific investigations.
- Design a scientific investigation.
- Develop descriptions, explanations, predictions, and models using evidence.
- Think critically and logically to identify the relationships between evidence and explanations.
- Recognize and analyze alternative explanations and predictions.

Threads of Evidence

Summary

In this activity students examine hairs and fibers under a microscope to look for identifying characteristics. Comparing known and unknown characteristics for matches and drawing conclusions based on observations will help students refine their understanding of scientific inquiry as they develop scientific investigation skills. Check out **Reproducibles,** page 21, for identifying hair and fiber characteristics.

Materials

(per student group)

- Slides with sample hairs and fibers (see note)
- Compound microscopes
- **Reproducibles,** page 21

Note: You or your students can collect a variety of samples from home and around the classroom; some hairs with roots, cotton thread, wool, artificial fibers (such as carpet or fleece), and animal hairs. Before beginning this activity, coat the samples with clear nail polish to keep them in place on the slides. These slides will be the known characteristics and should be clearly labeled.

NSES

- Different questions suggest different kinds of scientific investigations. Some involve observing and describing; some involve collecting specimens; some involve experiments; some involve seeking more information; some involve discovery; and some involve making models.
- Scientific investigations sometimes result in new ideas and phenomena for study, generate new methods or procedures for investigations, or develop new technologies to improve the collection of data.

Background

Found at many crime scenes, hairs and fibers provide valuable clues for helping establish identity of a suspect and that person's whereabouts. Investigators often collect these "threads of evidence" using small vacuum cleaners with special filters.

Hairs and fibers are everywhere; it is impossible for a person to move about without picking up or shedding small fibers and hairs. Fibers found on victims led police to arrest criminal Wayne Williams in Atlanta, Georgia, in 1979. The forensic investigators followed these threads of evidence to Williams's apartment and car, where they found carpet fibers identical to those found on his victims.

What to Do

1. Ask students how they would identify hairs. What are the identifying features of hairs and fibers? Is there a difference between human hair and animal hair?
2. Give each student group a set of glass microscope slides prepared in advance with hairs and fibers.
3. Instruct the students to use their microscopes to make careful observations of each known sample. Encourage them to use information in **Reproducibles,** page 21, to help identify characteristics of each sample.
4. Give each pair a sample with unknown characteristics and challenge students to identify the unknown by analyzing it and comparing it to the known.
5. Invite students to form a hypothesis explaining how the unknown sample may have been left at the crime scene.
6. Conclude by sharing with students that critical observation and analysis of hairs and fibers can provide the evidence forensic scientists use to solve a case by ruling out a suspect or making a rough match.

Extension

Have students handle several materials, such as fur, a tweed-fiber coat, or a piece of carpet. Ask them to drag the samples over a sleeve or pant leg, using clear tape to collect any deposited fibers. Tell the students to transfer the samples to glass slides and coat them with clear nail polish. They can examine the samples under the microscope and compare them to the sample slides. Students will learn how easily fibers are attracted to clothing. ■

A Activities

Help your students engage in inquiry-based learning

Discovery Channel School

If the Shoe Fits

Summary

In this activity students form a cast of a shoe print in moist sand or plaster of Paris. Students make a scale drawing of the shoe impression and use observation and mathematical skills to record details and to offer a hypothesis to explain their ideas. Students learn how to calculate estimated height based on the length of the footprint.

Materials
(per student group)
- 3-inch-wide strips of poster board long enough to encircle the print
- rubber dish pans or similar container half-filled with moist sand or soil
- chicken wire cut to fit over shoe impressions
- plaster of Paris
- old toothbrushes
- shellac or hair spray
- petroleum jelly
- measuring tape

Note: This activity can be done entirely in moist sand instead of in plaster of Paris.

NSES
- Different questions suggest different kinds of scientific investigations. Some involve observing and describing; some involve collecting specimens; some involve experiments; some involve seeking more information; some involve discovery; and some involve making models.
- Mathematics is important in all aspects of scientific inquiry.

Background

Shoe prints can provide clues. Forensic scientists look for two levels of matches. One is matching the print to the shoe's manufacturer; the other is finding accidental characteristics. A forensic scientist can estimate a person's height by using a ratio. Foot size is generally about 15 percent of a person's total height. Anthropologists and archaeologists also use ratios to calculate the size of animals and humans based on their remains.

What to Do

1. Engage students in brainstorming a list of clues you might find from discovering a shoe print. What tools would you use to gather your evidence?
2. Guide each group in the following steps to make a plaster cast.
   - Invite a volunteer to make an impression with a shoe in the sand tray.
   - Spray the impression with shellac or hairspray to ensure an easy release of the plaster mold from the sand.
   - Encircle the print with the poster board strips.
   - Coat the inside of the strip with petroleum jelly so the plaster will release easily.
   - Mix plaster of Paris (or moist sand) and coat the print with half of the mixture. Cover the plaster with chicken wire for stability.
   - Pour enough plaster to cover the wire.
   - Allow the mold to set overnight.
   - Clean the mold using an old toothbrush.
   - Label the back of the mold with information about shoe size, brand, and stance when the mold was taken.
3. Ask the groups to make a 1:1 scale diagram of the shoe impression on graph paper. Encourage students to identify and label print features.
4. Discuss the print. Are there any accidental characteristics that make this shoe unique to the wearer? Does the tread have a crack in it or unusual wear in certain areas?
5. Challenge the students to estimate the height of the person who left the footprint. Tell them that a person's foot size is approximately 15 percent of their height. Remind the students that shoe prints will be larger than actual foot size. Have them compare the calculated height to the actual height.
6. Go back to the brainstorming list from Step 1. Engage students in refining the list based on their experiences in this activity.

Extension

Sometimes soil samples are embedded in shoe treads or tires. Collect sand samples around school to compare with a known sample found embedded in the shoe print. Allow students to examine the samples with a hand lens to find out where a suspect may have been before arriving at the crime scene. Read "The Invasion of the Fire Balloons" (**Interdisciplinary,** page 12) to learn more about soil samples. ■

Discovery CHANNEL SCHOOL

Bibliography

Great books for content, leisure reading, or classroom reference

Books for teachers

Beyond the Crime Lab: The New Science of Investigation, rev. ed.
Zonderman, Jon. John Wiley & Sons, 1998.

This book explores recent crimes to look at the most up-to-date methods of forensic research. It details how forensic science is employed in every step from crime scene investigation to laboratory examination. Among the cases covered are the Unabomber, TWA Flight 800, and the Oklahoma City bombing.

The Casebook of Forensic Detection: How Science Solved 100 of the World's Most Baffling Crimes
Evans, Colin. John Wiley & Sons, 1998.

A useful and entertaining text that describes the scientific methods—from fingerprints to ballistics—that have helped solve famous crimes. Each section introduces a forensics area, its history, and its pioneers. Though not as technically detailed as other works in the field, this book is a good introduction.

Crime Science: Methods of Forensic Detection
Nickell, Joe and John F. Fisher. University of Kentucky, 1998.

This book gives a broad overview of forensic science featuring case studies to illustrate key concepts. Extensive references for further reading follow each chapter.

Criminalistics: An Introduction to Forensic Science
Saferstein, Richard. Prentice-Hall, 1997.

A textbook that covers techniques, skills, and restrictions of modern forensic laboratories, as well as the role science plays in the criminal justice system. The book is for people with limited knowledge of the forensic sciences. ■

Books for students

Detective Science: 40 Crime-Solving, Case-Breaking, Crook-Catching Activities for Kids
Wiese, Jim. John Wiley & Sons, 1996.

The book is divided into five areas: crime-scene observation; evidence gathering; and analysis of chemical, biological, and documentary clues. Each chapter outlines simple experiments for children, although some require adult help.

Eyewitness: Crime and Detection
Lane, Brian. DK Publishing, 2000.

Using a popular lexigraphic format of color photos and illustrations, cutaway and cross-section illustrations, 3-D models, and maps, this reference book is a far cry from the dry connotations of that term. The text covers historic crimes and presents information about recent advances in forensic science.

Fingerprints and Talking Bones: How Real-Life Crimes are Solved
Jones, Charlotte Foltz. Yearling Books, 1999.

Students will enjoy this look at the science behind crime solving. Included are a glossary, cool crime facts, and a bibliography.

Forensic Science: Evidence, Clues, and Investigation
Campbell, Andrea. Chelsea House, 1999.

This book by a forensic reconstruction sculptor follows the process of crime solving from the moment of discovery through evidence gathering. She explains the way evidence is used.

Two-Minute Mysteries
Sobol, Donald J. Scholastic, Inc., 1991.

Short but exciting mysteries are told in a simple fashion that makes this book ideal for kids ages 9–11. The reader is given clues and has to "study" the evidence to solve crimes.

Solving Crimes: Pioneers of Forensic Science
Fridell, Ron. Franklin Watts, 2000.

Students interested in forensic science will enjoy the biographies of six men responsible for significant advances in the field.

The Wildlife Detectives: How Forensic Scientists Fight Crimes Against Nature
Jackson, Donna M. Houghton Mifflin, 2000.

A look at wildlife forensic scientists. Jackson explains how detectives interpret clues, employ electronic equipment, and follow tips to catch a criminal. Included are "Wild Files" that explain wildlife law, careers in animal crime labs, and efforts to protect endangered species. Also featured are engaging full-color photographs, a list of follow-up suggestions, and a glossary. ■

Careers

Share this section with students to help them plan for higher education and careers

Occupations Relating to Forensics

Encourage your students to contact experts in forensics-related fields of business for a classroom visit or to find question-and-answer sites on the Internet.

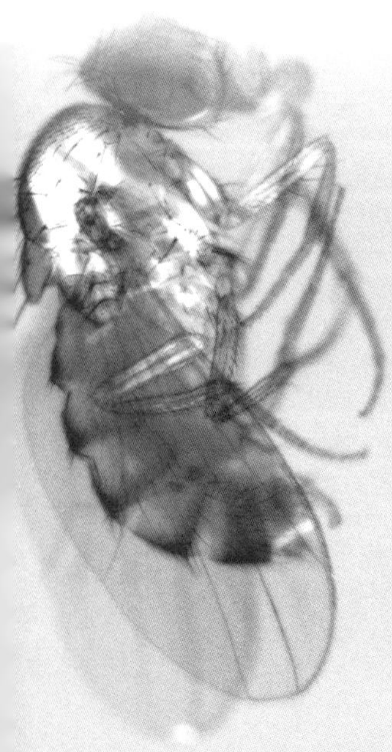

Forensic science is an ever-growing field. Most career opportunities in the field require at least a bachelor's degree in a physical science such as chemistry, biology, or physics. Coursework should include concentrations in chemistry, microscopy, and statistics. Some universities across the country now offer majors in forensic science or criminal justice programs—some offer graduate degrees; others have certification courses.

Below are some jobs you might want to investigate if you are interested in a career in forensic science.

Fingerprint Examiner

This laboratory job requires some coursework in fingerprint examination, evidence classification, and police photography. Fingerprint examiners analyze evidence using chemicals, photographs, and laboratory equipment such as lasers and fingerprint computers. They need an associate's degree or course work in criminology, as well as on-the-job training in latent fingerprint comparison, photography, and police identification. Additional course work in forensic science, chemistry, or biology may be required.

Firearms Examiner

A combination of crime scene and laboratory work, this job does not require an advanced science degree. Many examiners without degrees are police officers who transferred into the field and received firearms-identification training from their departments. The job primarily involves comparing ammunition to the firearm from which it was fired. Also, firearm examiners test firearm functionality and examine clothing and other items for gunshot residues. Examiners also determine manufacturers of ammunition components, as well as firearms that may have fired a particular bullet or cartridge case.

Forensic Entomologist

If you think insects are interesting, consider this career. Forensic entomologists study insects and other invertebrates in a legal context. Entomologists examine bugs on a corpse to determine a time of death and insect remains on evidence that may help solve a crime.

Forensic Nurse

A growing field for people interested in caring for victims of violent crime. This career involves physical care, psychological counseling, and testifying in court about the state of a victim's physical and mental health.

Forensic Odontologist

Dentists who use their knowledge of bite marks to help solve crimes are forensic odontologists. They also make identifications of bodies through dental charts, tissue samples, and computer analysis.

Questioned-Document Examiner

A questioned-document examiner, or QDE, inspects wills and other documents to certify that they are genuine. Examining details in a questioned sample and comparing them to a sample of known authorship often determine the authorship of a written document. A QDE's main responsibility is handwriting analysis, examining the design, shape, and structure of a handwriting sample. QDEs look at letter formations, connecting strokes between the letters, upstrokes, retraces, downstrokes, spacing, baseline, curves, size, and other characteristics. A QDE requires an associate's degree or coursework in criminology with an emphasis in handwriting analysis.

Trace Evidence Examiner

In the laboratories of law enforcement agencies, far from crime scenes, these examiners test collected evidence such as glass and paint objects, explosives, and fire debris. In addition, examiners often testify in court about their findings. A trace evidence examiner must have a background in chemistry, with specific knowledge of how to analyze fibers, soil, metals, building materials, and other small evidence. In some cases, trace evidence examiners may analyze hairs, fibers, body fluids, fingerprints, gunshot residue, questioned documents, controlled substances, plastics, tool marks, and other forensic trace evidence submitted in scientific criminal investigations. Those working in this job usually have an advanced degree in criminology, forensic science, chemistry, biology, or physics. A bachelor's degree in any of the above fields and several years of full-time experience in forensic science work meet the requirements of many laboratories. ■

Discovery CHANNEL SCHOOL

Demonstrations

The excitement's up front with you, the teacher, for these classroom "events"

D

NSES

- Scientific investigations sometimes result in new ideas and phenomena for study, generate new methods of procedures of investigations, or develop new technologies to improve the collection of data.
- Different questions suggest different kinds of scientific investigations. Some involve observing and describing objects, organisms or events; some involve collecting specimens, some involve experiments, some involve seeking more information; some involve discovery of new objects and phenomena; and some involve making models.

General Demonstration Guidelines

For safety or materials reasons, demonstrations need teacher direction. But they can still provide students with opportunities to develop abilities and understandings of inquiry, especially when the demonstration replicates a "great discovery."
Questioning strategies should focus on developing the following abilities of inquiry:

- Design a scientific investigation.
- Develop descriptions, explanations, predictions, and models using evidence.
- Think critically and logically to identify relationships between evidence and explanations.
- Recognize and analyze alternative explanations and predictions.

Discovering Invisible Evidence

Summary

Students participate in a simulation of collecting evidence at a crime scene. They learn one way investigators find "invisible" evidence.

Background

Investigators use luminol, a mixture that causes blood and other stains no longer visible to the naked eye to fluoresce blue when exposed to UV light.

Materials

- light stick
- old dark T-shirt
- spray bottle filled with water (to simulate luminol spray)
- objects such as books, candlesticks, and a water glass

What to Do

1. Before class, activate a light stick by removing it from the foil wrapper and bending it until you hear the snap. Wear goggles and gloves while cutting the light stick with a sharp knife. Allow some of the fluid that simulates a bloodstain to dribble onto the T-shirt and arrange the shirt so that the glowing spot is hidden. Arrange the T-shirt and several other objects on a table to simulate a crime scene.
2. Direct students' attention to the scene. Ask them to suggest how investigators might approach such a scene: Would investigators analyze the scene in a particular order? Is all the evidence visible when investigators first arrive on the scene? What kinds of evidence might be hard to detect?
3. Ask a student volunteer to put on latex gloves and examine the evidence by first spraying the area lightly with the simulated luminol solution. Darken the room and instruct the student to pick up the T-shirt. The spot with the fluid from the light stick will glow blue, indicating a non-visible stain. Conclude by asking students to speculate about other types of invisible evidence. ■

Forensics Lab

Summary

Students learn how to use a systematic examination of evidence to analyze clues left by a hunter found frozen in the Italian Alps 5,300 years ago. Archaeologists use forensic science to uncover the mysteries of what this man was like and what may have happened to him.

Background

Biological samples must be carefully collected and analyzed. Building the scenario of what might have taken place requires careful examination, making matches, and analyzing distinctive properties and characteristics.

What to Do

1. Set up an area of the classroom with props that suggest an ice field with a model of Ice Man's mummified body. Add artifacts that were found with Ice Man. (See **Unsolved Mysteries,** page 25.)
2. Tell the students that you've created a model of Ice Man's discovery site in the Italian Alps. Have students record what they see at the discovery site in the form of charts, diagrams, and photographs. (See **Reproducibles,** page 21.) This is the first step investigators take after the site or crime scene is secure.
3. Guide students to collect samples, using the following steps and precautions not to contaminate the site.
   - Put on gloves.
   - Use forceps to put the evidence in carefully labeled plastic bags.
   - Close and label plastic bags.
4. Invite students to "sign out" evidence bags, analyze the evidence, and make a hypothesis in the form of a scenario describing what they think may have happened to Ice Man or what they think his life was like. ■

Experiments

To provide students with guided practice in inquiry

General Experiment Guidelines

The teacher should provide the skeleton of the experiment, but students still should be involved in making decisions regarding data collection, controlling variables, and depth of study. The aim of the experiments should be to focus students on these abilities:

- Design (or modify) a scientific investigation.
- Use appropriate tools and techniques to gather, analyze, and interpret data.
- Develop descriptions, explanations, predictions, and models using evidence.
- Think critically and logically to identify the relationships between evidence and explanations.
- Recognize and analyze alternative explanations and predictions.
- Communicate procedures and explanations.
- Use mathematics.

Identifying a Mysterious Substance

Summary

Students describe and analyze chemical and physical properties of several white powders in order to match a known substance. They are encouraged to vary the mixtures and look for a combination of reactions that match the reactions and characteristics of the unknown white powder.

Materials

(per student group)

- watch glass, petri dish, or reaction plate
- test tube or small jar
- hand lens or compound microscope
- white powders in small vials labeled A–E
- vinegar in a small dropper bottle
- iodine in small dropper bottle
- water
- water dropper or beryl pipette
- litmus paper or universal indicator

NSES

- Different questions suggest different kinds of scientific investigations. Some involve observing and describing; some involve collecting specimens; some involve experiments; some involve seeking more information; some involve discovery and some involve making models.
- Scientific investigations sometimes result in new ideas and phenomena for study, generate new methods or procedures for investigations, or develop new technologies to improve the collection of data.

Background

Preparation: Make one set of known powders and an evidence bag with a mystery powder for each group of students. Powders that work well are flour, baking soda, salt, Epsom salts, sugar, cornstarch, and baby powder. Use letters, as students should identify the powders by their properties and not by their household names. Place the "known" powders in marked vials that students will use to compare with the mystery powder. Place a dry mixture of two of these powders in a small plastic bag marked "evidence."

What to Do

1. Begin by showing students an evidence bag of white powder. Tell a story appropriate for their age and experiences about finding this powder at the scene of a crime. Let them know that their job is to figure out what the powder might be by comparing it to known substances.
2. Distribute group sets of known powders. Ask students to suggest types of tests that they can do in the classroom to identify each powder. List their ideas, suggesting these if needed: Test for starch with a drop of iodine, test for reaction with vinegar, test for solubility with water, and test for pH with litmus paper.
3. Involve students in setting up a data table for this examination. The data table needs to include space for a sketch of the powder; physical description such as color, texture, shape of crystal or particle; and space for the results of the four tests described in Step 2.
4. Invite students to begin their tests with a close examination of each powder under the microscope or hand lens. Challenge them to note the shapes of individual grains or crystals. Allow time for them to perform and record results for the remaining tests.
5. Give each student group an evidence bag of the unknown white powder. Challenge students to identify the contents. Invite them to repeat all the tests on this unknown substance. Remind them to use the data table to help them analyze the results of their experiments on the unknown and determine combinations of powders used in the unknown.
6. Ask groups to prepare a statement that uses evidence gained from their tests to defend any conclusion reached as to the substance's identity.
7. Conclude by having them suggest what other kinds of methods might be useful in analyzing substances. ■

Testing a Document's Authenticity

Summary

This experiment uses chromatography to determine which ink was used to write a document. Paper chromatography is used to separate ink mixtures. Water carries the pigment up the paper by capillary action, an attractive force between the water molecules and the paper fibers. The ink separates into a rainbow pattern, but not all black inks leave the same patterns. Investigators may use individual color patterns to identify a particular brand of pen.

Materials

(per student group)

- 6 different water-soluble black markers
- 10 precut strips of coffee-filter paper (cut into strips 10 × 3 cm each)
- 6 clear plastic cups or beakers
- clear tape
- rulers
- hand lenses
- stopwatch
- mystery document written with one of the 6 markers

Background

Investigators use chromatography to match unknown substances with known substances. With the use of a mass spectrometer, a device that measures light absorption, investigators take chemical "fingerprints." These fingerprints help investigators identify the substances. If they find an exact match, investigators try to locate the manufacturer.

What to Do

1. Ask students to think about ways to determine a document's authenticity. What can you determine about the paper? What can you find out about the ink? Are there unique identifying marks that could help determine the writer of the document? Could there be fingerprints on the paper? What kind of pen might have been used to write the document? What can you tell about the ink? Are all black inks the same? If the ink is a mixture of pigments, how can we separate the pigments?
2. Demonstrate how black ink separates into several pigments by using chromatography. Make a black dot three centimeters from the end of a strip of coffee-filter paper. Fill a clear plastic cup with 2 centimeters of water and then lower the coffee-filter strip so that it touches the water but doesn't reach the black dot. The water will travel up the strip and deposit the color pigments by capillary action. Show the students the chromatography strip and tell them this is a sample of the ink from the document. Ask them to design an experiment using chromatography that will help them identify the type of pen used to write the document.
3. Show the students the mystery document. Create a story about the document. Ask the students to design an experiment using chromatography that will help them identify the type of pen used to write the document.
4. Encourage students to consider what kinds of measurements they should record. Suggest that they use a ruler to make measurements and a stopwatch to record the length of time it takes for the different colors to move up the coffee-filter strips.
5. Ask students to design a data table that will help them keep track of their results. It will need to include space for each filter strip sample and important information about the pen.
6. Set out the six water-soluble black pens, plastic cups, tape, rulers, precut coffee-filter strips, and stopwatches. Invite a student from each group to collect materials. Allow time for groups to conduct tests on ink from all six pens.
7. After groups have collected data about the six ink samples, distribute a chromatography strip of the ink taken from the sample document. Ask the student groups to examine the strips carefully and determine which one of the strips matches the document ink. ■

Free Stuff

Charge up your Forensics unit—at no extra charge!

DISCOVERY CHANNEL SCHOOL

The Best of the Web

Handbook of Forensic Services
www.fbi.gov/programs/lab/handbook/intro.htm

An online FBI publication that contains information about all aspects of forensics. The text is technical in areas, but the content is excellent for any related research project.

The Killer's Trail
www.pbs.org/wgbh/nova/sheppard/

Sponsored by Nova Television series on the Public Broadcasting System, this site presents the case of Dr. Sam Sheppard, a murder case that was the basis of *The Fugitive* TV series and movie. Among the links for the site are pages on DNA fingerprinting and a computer-generated 3-D sketch of a possible suspect.

Evidence: The True Witness
http://library.thinkquest.org/17049/gather/

This site presents an extensive look at forensic science, created as part of the ThinkQuest competition, an international contest for secondary school students to create online learning tools. Students from the United States and Australia created this award-winning site for the 1998 competition. Included is a forensics game that requires critical thinking skills.

U.S. Fish and Wildlife Service Forensics Laboratory
www.lab.fws.gov/

Originated from the nation's first wildlife forensics laboratory, this site offers numerous links that explain how forensic science is applied to animal-related crimes. ■

Posters, Pamphlets, and Programs

Junior Special Agent Program

This school program for fifth- and sixth-graders is sponsored by the FBI as part of the agency's Community Outreach Program. The program consists of classroom activities, fitness training, and guest appearances by FBI agents who specialize in white-collar crime, computer fraud, and terrorism investigations. In addition, agents supervise field trips and assist in mentoring and tutoring programs for students in some subject areas. Obtain information through your local FBI office.

U.S. Fish and Wildlife Service Forensics Lab
www.lab.fws.gov/lab/docs/idnotes/idnotes.htm

Online Identification Kit allows viewers to see the main characteristics that guide forensic wildlife examiners. This thorough kit has precise drawings of reptiles, birds, and mammals, as well as a list of specific identifying characteristics. A downloadable application is available for viewing the pages. ■

Games and Puzzles ☻

Keep the learning going while everyone has some fun

R U Clueless?

See answers on page 32.

See if you have what it takes to be an eagle-eye investigator. Find all the forensic terms in the word search below and crack the case! *Hint:* Look forward, backward, horizontally, vertically, and diagonally.

AUTOPSY	LATENT
BALLISTIC	ODONTOLOGY
BLOOD	OFFENDER
CLUE	PATHOLOGY
EVIDENCE	PROFILE
FIBER	REAGENT
FINGERPRINTS	SCENE
FORENSIC	STRIATIONS
GENE	TOXICOLOGY

S	K	T	T	F	Z	O	E	P	Y	X	S	Z	L	B
F	T	Z	Y	R	O	C	F	G	R	T	L	A	W	U
R	J	R	Q	G	N	R	O	F	N	O	T	X	F	S
X	E	N	I	E	O	L	E	I	E	E	F	N	Q	T
F	X	A	D	A	O	L	R	N	N	N	F	I	C	U
F	Q	I	G	T	T	P	O	T	S	I	D	V	L	Y
Y	V	H	N	E	R	I	U	C	B	I	R	E	W	E
E	D	O	N	E	N	W	O	E	I	Q	C	E	R	S
M	D	I	G	X	L	T	R	N	B	X	S	D	Y	W
O	X	N	D	O	O	L	B	I	S	B	O	A	S	V
Q	I	P	A	T	H	O	L	O	G	Y	D	T	P	E
F	C	L	U	E	W	G	E	N	E	C	R	W	O	N
K	U	Y	I	J	S	U	A	Y	M	I	E	P	T	E
B	A	L	L	I	S	T	I	C	T	J	X	E	U	C
V	R	K	M	P	O	K	S	M	O	U	L	T	A	S

This word-search game was created using **Puzzlemaker** at **http://puzzlemaker.school.discovery.com.** Students may visit the Web site and make games of their own. ■

Who Ya Gonna Call?

Forensic scientists range from insect experts to specimen specialists. On the left below are jobs in different fields of forensic science. On the right are descriptions of crime scenes. Match up each job title to the correct scene.

a. odontologist
b. entomologist
c. forensic anthropologist
d. serologist
e. questioned-document examiner
f. pathologist
g. ballistics specialist
h. toxicologist

_____**1.** A body gives off the smell of almonds—the same smell as deadly cyanide.

_____**2.** A shipment of drugs from an unknown area is found with an unusual moth cocoon inside.

_____**3.** A body with bruises on the neck is pulled from a lake. Did the victim drown or was he strangled?

_____**4.** A robber leaves a half-eaten apple with unusual tooth marks at the crime scene.

_____**5.** An unidentified skeleton is found in a remote wilderness area.

_____**6.** Blood dots the wall at a murder scene and the man in custody has blood on his shirt.

_____**7.** A bullet lodged in the wall resembles a bullet taken from a victim.

_____**8.** Two credit card receipts show the same card but slightly different signatures. ■

Homework Helpers

Build a variety of assignments around this information

Life Cycle of a Fly

Forensic entomologists use their knowledge about insects to help answer questions. Some species of flies are used as "clocks" because they have distinct stages of development. Each stage takes a particular length of time. The fly's life cycle starts with laying the eggs on an animal's carcass. The eggs hatch into tiny larvae called maggots. Fly larvae develop hard outer shells called pupal cases. The adult fly emerges from the pupal case and then the cycle starts all over again when the adult fly looks for a place to lay its eggs. Insects are cold-blooded, so forensic entomologists must know the time length for each of these stages in warm and cool weather. Colder weather can delay insects' life cycles.

In the following case investigators must establish the time of death for a deer found in the woods. The deer's carcass with its antlers removed was found 35 days after the hunting season started. The insects found on the carcass were the adult cheese skippers Piophilidae and the pupal cases of the blowfly *Calliphora vicina.* The daily average temperature was 77°F (25°C).

Use the charts and diagrams on this page to see if you can calculate how long ago the deer was shot. Was it shot during hunting season or before?

Answers are on page 32. ■

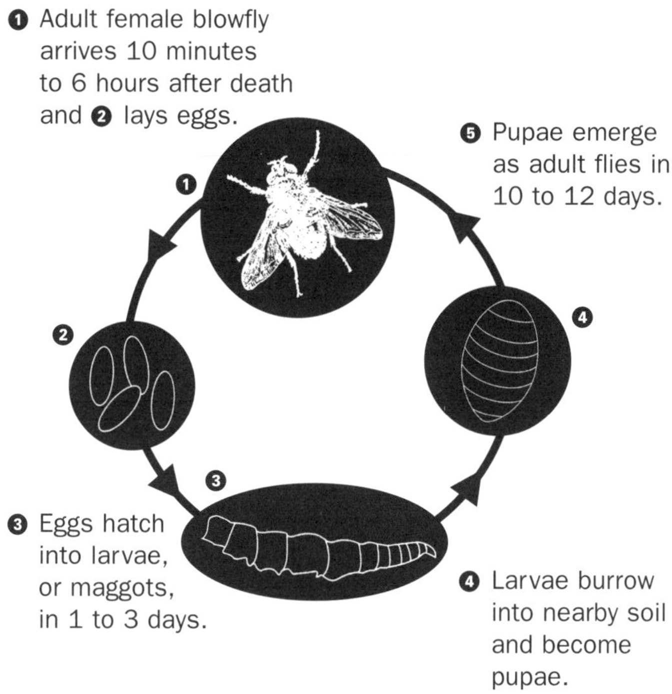

Number of Days Required for Insects to Reach Adult Stage*

Insect	5	10	15	20	25	30	35	40	45	50
Blowfly—*Calliphora vicina*				adult						
Cheese skippers—Piophilidae								adult		
Lesser houseflies—Fannidae										adult

*Based on an average daily temperature of 76°F (23°C)

Interdisciplinary

Connections to other subjects add new dimensions to science

National Science Education Standards

- The program of study in science for all students should be developmentally appropriate, interesting, and relevant; emphasize student understanding through inquiry; and be connected with other school subjects.
- Mathematics is important in all aspects of scientific inquiry.
- Science influences society through its knowledge and world view.

Language Arts: Create a Crime Scenario

Mystery authors have been weaving tales of crime and intrigue that have captivated readers for centuries. Sir Arthur Conan Doyle, author of *The Adventures of Sherlock Holmes,* crafted amazing stories with remarkable characters. Mr. Holmes and Dr. Watson left no stone unturned as they solved crimes in 19th-century England. Have students read a Sherlock Holmes adventure and discuss how the author let the story unfold, how the characters analyzed evidence, and how deductive reasoning solved the case. Suggest *Six Great Sherlock Homes Stories* by Arthur Conan Doyle (Dover, 1992). Have your students write a crime scenario weaving all the clues to tell their own story of "who done it" and why. ■

Math: Scale Drawing of a Crime Scene

Encourage students to use their mathematical skills to make a scale drawing of a crime scene. Have the students measure the outer perimeter of the crime scene. Ask them to decide how the drawing will best fit on graph paper. Divide the largest measurement by the number of squares on the graph paper to get a ratio. For example, one foot equals one square on the graph paper. With a strip of poster board students can make a scale and use it to measure distances that are not square. The drawings should be clearly labeled and include a legend with a scale. ■

Interdisciplinary

Connections to other subjects add new dimensions to science

National Science Education Standards

- The program of study in science for all students should be developmentally appropriate, interesting, and relevant; emphasize student understanding through inquiry; and be connected with other school subjects.
- Mathematics is important in all aspects of scientific inquiry.
- Science influences society through its knowledge and world view.

Art: Painting Forgery

Hans van Meegeren needed to prove that he had forged a painting. He was on trial for collaboration with the Nazis. Why? He sold a painting to Hermann Goering, commander in chief of the German air force and a powerful leader in Nazi Germany, for $2 million during World War II. Van Meegeren had done such a good job of forging a painting by 17th-century Dutch artist Jan Vermeer that he had a hard time persuading forensic scientists that the painting was a fake. They concluded that it was an original Vermeer and insisted that Van Meegeren was simply trying to save his life.

Authorities decided to reexamine the painting. Van Meegeren disclosed the details of how he spent four years perfecting his forgery. He scraped paint off 17th-century paintings, then painted over the surface. Van Meegeren baked the paintings in an oven and rolled them in a metal tube. He was careful to use paints that had the same chemical makeup as in Vermeer's time. But he was not aware that the paint manufacturer had changed a blue paint. Under the microscope the forensic scientist saw diamond-shaped crystals indicating cobalt blue, a paint that was not available during Vermeer's time.

Have your students make paint swatches of different oil and acrylic paints to look at under the microscope. Ask them to look for identifying characteristics that might help a forensic scientist match the paint. Encourage your students to look at paintings by the Dutch Masters. ■

History and Geology: The Invasion of the Fire Balloons

Following the evidence trail can take an investigator from a microscopic view to a global view. The case of "The Invasion of the Fire Balloons" all started with grains of sand and ended up on a beach in Japan. During World War II, when Japan was at war with the United States, the Japanese launched a fire balloon attack from one of their northern island beaches. The balloons were designed to drift across the Pacific Ocean and land on the U.S. mainland. In January 1945, mysterious balloons were sighted from California to Alaska, with some as far inland as Wyoming. Observers witnessed the balloons exploding, sometimes creating flames as high as 30 feet. Forensic geologists, who study changes in the landscape, analyzed the sand ballast at the landing sites. Minerals and microscopic organisms in the sand led investigators to a beach in Japan where the balloons had been launched. More than 500 balloons reached North America before the launch site was located and destroyed.

Have your students examine sand samples from various locations. Ask them to sketch their samples in detail and analyze the components. Students should expect to find fine grains composed of silica from beaches and stream beds. Silica (SiO_2) appears angular under a microscope. Other sands may have larger particles, such as sands generated from erosion and weathering of granites, which are found in arroyos in the desert. Another type of sand, gypsum, is found in White Sands, New Mexico. Sands and soils tell a story about the environment of a certain area and give clues to both past and present geology. Use the Internet to connect with other schools in different parts of the country and ask them to do a sand exchange and swap sand samples with your school. ■

Just for Fun☻

A little lightheartedness leavens any learning!

Quirky Courtroom Quotes

Save these for a rainy day—or anytime.

The following are actual courtroom conversations reported in the Massachusetts Bar Association *Law Journal.*

The questions below were asked by a defense attorney; the answers were from a forensic pathologist.

Q: Doctor, before you performed the autopsy, did you check for a pulse?
A: No.
Q: Did you check for breathing?
A: No.
Q: So, was it possible that the patient was alive when you began the autopsy?
A: No.
Q: How can you be sure, doctor?
A: Because his brain was sitting on my desk in a jar.
Q: But could the patient have been alive nevertheless?
A: Yes, it is possible he could have been alive and practicing law.

Questions asked by a prosecutor:

"Were you alone at the time, or by yourself?"

"Were you present when your picture was taken?"

Questions from a defense attorney:

"Did he kill you?"

"You were there until the time you left, correct?" ■

Sherlock Holmes Gets It Wrong

Of all the great detectives in literature, perhaps none is so admired as Sherlock Holmes. The fictional sleuth created by Sir Arthur Conan Doyle appeared in 60 stories between 1887 and 1927. One of the first and most popular of the Sherlock Holmes stories was *The Adventure of the Speckled Band* published in 1892. In this tale, two victims are found dead inside a locked bedroom with no visible marks on their bodies.

Holmes solves the crime by determining that the murderer, Dr. Roylott, keeps a poisonous "speckled viper" in a room next to that in which the victims slept. Before the victims go to bed, Roylott places a bowl of milk under the bed. Then, in the middle of the night, Roylott blows a high-pitched whistle, a signal to which the snake has been trained to respond. The deadly reptile climbs up a bell cord, through a vent connecting the rooms, and down a bell cord seeking the milk that it receives as a reward for climbing. Once it encounters a sleeping victim, it bites and kills with its venom, leaving almost invisible fang marks.

This tale was made into a movie in the 1930s. However, modern herpetologists—reptile experts—have pointed out many flaws in the plot. First, snakes are deaf and cannot be trained. Second, snakes don't drink milk. Third, a "viper" is unable to climb up or down a rope. The only snakes that can do that are in the constrictor family—and they are not poisonous!

As Holmes might say to his assistant: "It's elementary, Watson. Sir Arthur was a poor sleuth regarding the truth about a venomous viper." ■

Crime Rhyme

A detective who testifies in court tells the "sleuth truth." See if you can figure out the rhyming words for the forensic riddles that complete each sentence below.

1. An expert in wildlife crime specializes in ______ ______.
2. A bullet hole could also be called a ______ ______.
3. A forensic odontologist trying to match teeth marks to the correct suspect is looking for the ______ ______.
4. A detective searching for a clue by studying a mark left at a crime scene is looking for a ______ ______.
5. A forensic expert with knowledge about illegal substances and insects at crime scenes specializes in ______ and ______.
6. The head of a college department that offers courses in genetic fingerprinting is the ______ ______.
7. A terrific crime laboratory is a ______ ______.
8. A footprint at a crime scene is a ______ ______.
9. Arson investigators who study fires that are set by criminals are looking for ______ ______.
10. A man who gives polygraph examinations to other people is a ______ ______.
11. When you select evidence that will be used in court, you ______ ______.
12. A police chief is the ______ ______.

Answers are on page 32. ■

K Key Concepts

Content standards to be addressed in teaching Forensics

NSES

These concepts are adapted from content standards for grades 5–8, pages 142–171, *National Science Education Standards.*

Science as Inquiry (p. 148)

- Different questions suggest different kinds of scientific investigations. Some involve observing and describing objects, organisms or events; some involve collecting specimens; some involve experiments; some involve seeking more information; some involve discovery of new objects and phenomena; and some involve making models.
- Scientific explanations emphasize evidence, have logically consistent arguments, and use scientific principles, models, and theories.
- Scientific investigations sometimes result in new ideas and phenomena for study, generate new methods or procedures for investigation, or develop new technologies to improve the collection of data. All of these can result in new investigations.
- Mathematics is important in all aspects of scientific inquiry.

Science and Technology in Society (pp. 169–170)

- Technology influences society through its products and processes. Technology influences the quality of life and the ways people act and interact. Technological changes are often accompanied by social, political, and economic changes that can be beneficial or detrimental to individuals and to society. Social needs, attitudes, and values influence the direction of technological development.
- Science and technology have advanced through the contributions of many different people in different cultures, at different times in history. Science and technology have contributed to economic growth and productivity among societies and groups within societies.
- Societal challenges often inspire questions for scientific research, and social priories often influence research priority through the availability of funding for research.
- Science cannot answer all questions and technology cannot solve all human problems or meet all human needs. Students should understand the difference between scientific and other questions. They should appreciate what science and technology can reasonably contribute to society and what they cannot do. For example, new technologies often will decrease some risks and increase others. ■

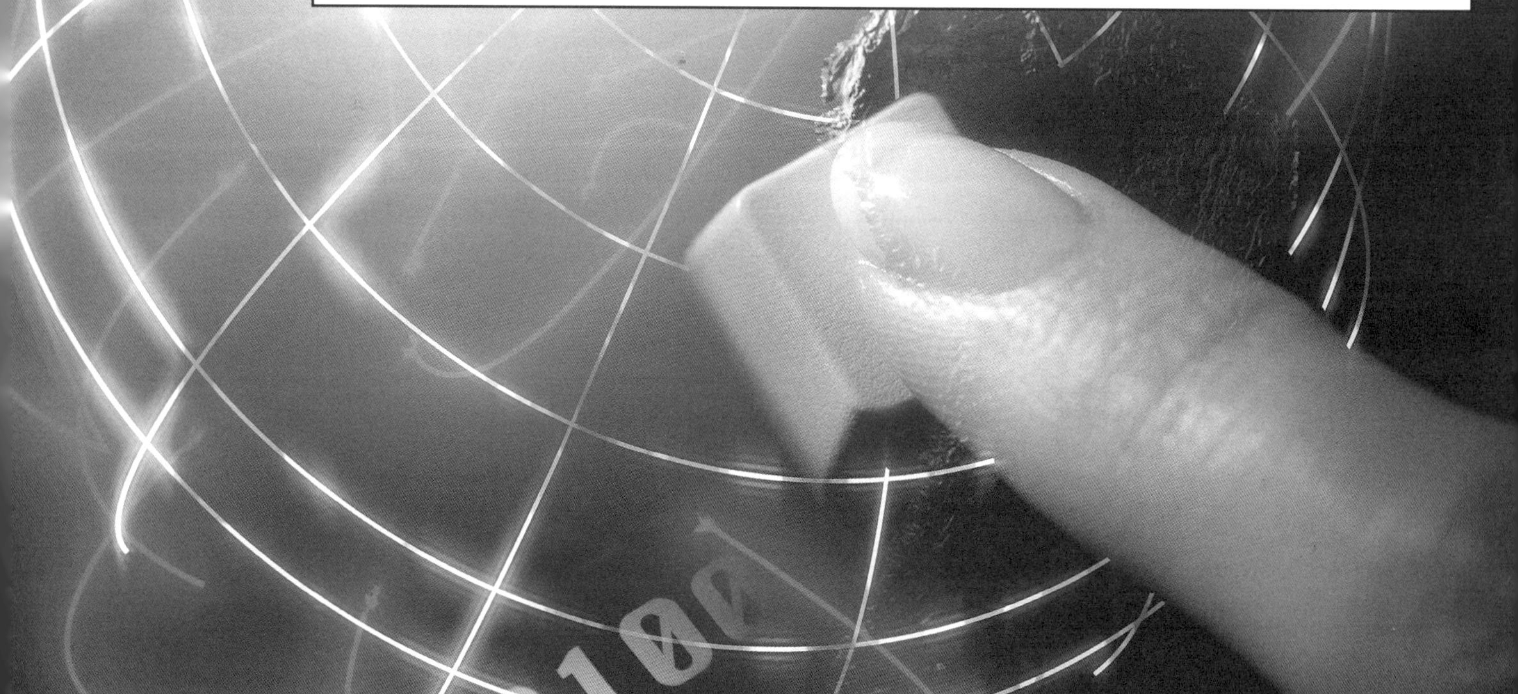

Leisure Activities

L

Reinforce the study of Forensics in your students' "off" hours

Virtual Sleuthing on the Web

Share these ideas with students.

Forensic Files

www.discoverlearning.com/forensic/docs/index.html

The Association for the Promotion and Advancement of Science Education (APASE), presents an interactive mystery adventure, an online game that asks kids to use their inferential reasoning skills to solve a nonviolent crime. Students can navigate through the game by clicking on icons, graphics, or text links.

The Case

www.mysterynet.com/thecase/

Mysterynet.com presents new mini-mysteries every week for online users to solve. This extensive archive of mysteries includes links to other interesting and educational forensic science Web sites for young people. ■

Movies

The Murders in the Rue Morgue

Several versions of this classic murder mystery may be available in your local library or in a well-stocked video store. The most recent version is a 1986 made-for-TV version starring George C. Scott as Auguste Dupin, one of literature's first detectives.

Clue

The movie version of this board game stars Lesley Anne Warren, John Astin, and other well-known actors. It's a comedy, but one that tests a viewer's powers of deduction. Available in most video stores and libraries.

The Thin Blue Line

This Academy Award-winning documentary details the case of a man wrongly convicted of the murder of a police officer. Based on actual evidence, director Errol Morris re-creates the crime and proves the innocence of the person convicted of the crime. ■

Are You Game?

The board game Clue offers deduction fun for kids. Players must solve a crime committed in a mansion by determining the suspect, the weapon, and the room in which the crime took place. Evidence is in the form of Clue cards set aside in a "top secret" envelope. Remaining Clue cards are passed out to players, who know that the cards they hold are not among those found at the crime scene. By moving around a game board through various rooms in the mansion, players have opportunities to ask each other to reveal a piece of evidence. The first player to solve the crime through process of elimination and deductive reasoning wins the game. ■

M Misconceptions

Common—and mistaken—ideas about Forensics

Discovery Channel School

Don't You Believe It!

Set the record straight with your students.

All techniques of evaluating evidence are acceptable.

Before a new forensic technique can be used, scientists must accept it as accurate and reliable. Known as the Frye Rule, this policy protects the court from being exposed to unreliable techniques. Under a Frye examination, a court decides if a particular technique is acceptable to the scientific community.

DNA-profiling techniques have passed the Frye examination. The first case in the United States that used DNA as evidence took place in 1987. Since then, more than 24,000 cases have used DNA analysis. In some cases defense attorneys have argued that the collection of DNA samples and the analysis of the profile has not been accurate. This is why investigators must carefully keep track of the chain of custody. (See **Questions & Answers,** page 20, for more about the chain of custody.)

DNA profiles will disintegrate over time.

Forensic scientists used DNA profiles to determine if the bones buried in 1918 in a mass grave belonged to the Romanovs, Russia's last royal family. Analyzed in 1992, the DNA in the children's bones was an exact replica of that in the segments of genetic code that can only be passed down through inheritance from the mother—proving that the DNA held up after three-quarters of a century.

After nearly two decades DNA evidence freed an imprisoned man in 1989. A. B. Butler spent 16 years in prison before officials tested his DNA. The test proved that his DNA was not at the crime scene. In many cases, however, DNA evidence gets lost or destroyed. The U.S. Congress is considering legislation that would require police to preserve DNA evidence as long as a convicted suspect remains in prison.

Law enforcement agencies are considering developing a database of DNA profiles of those convicted of murder and other major crimes. Some ordinary citizens may also want their DNA in a database for identification purposes should they be involved in a deadly disaster.

A lie detector detects lies.

A polygraph, or lie detector, records physiological changes such as increases in heart rate, breathing, and perspiration. The polygraph records these changes on a graph while the person answers a series of questions. Someone telling the truth usually will not show physiological changes. But if that person lies, measurable changes will show up. However, some people have trained themselves to remain calm, and some innocent people react as if they are guilty. For this reason, polygraphs have an accuracy rate of less than 95 percent. ■

Noteworthy People

Share these achievers' stories with your students

Dr. Henry Chang-Yu Lee

Students might like to research projects or make presentations on other achievers, including

- **Matthew Orfila (19th-century French criminologist)**
- **Henry Goddard (19th-century British forensic detective)**
- **Peter Neufeld (American DNA legal scholar)**
- **Sir Alec Jeffreys (British geneticist)**

December 22, 1938–

Thanks to the work of Henry Chang-Yu Lee, investigators can link a bloody fingerprint at a crime scene to a suspect. Dr. Lee is a pioneer in the field of forensic science.

As a police captain in Taipei, Taiwan, in the late 1950s, Lee used the most common method of the time to solve crimes: interrogating suspects until they confessed. But Lee found that too many cases remained unsolved due to a lack of scientific crime-fighting methods.

When Lee immigrated to the United States in 1965, he pursued an education that allowed him to mix science with police work. It wasn't easy for him. Arriving in New York City with less than $50 in his pocket, Lee worked as a waiter and martial arts instructor while attending college. Lee received his Ph.D. in human genetics and molecular biology in 1975, and he explored ways to connect science to law enforcement. At the time, the field of forensics included fingerprinting, handwriting analysis, ballistics, and autopsy.

In 1975 Lee, a professor at the University of New Haven, Connecticut, established the first forensic science program in the United States. Scientists and specialists working with Lee demonstrated how biochemistry could play a critical role in forensics through blood-group typing, human bone-enzyme typing, and DNA profiling.

During the 1980s Lee's colleagues and students developed new methods to extract DNA from samples, a technique to enhance bloody fingerprints, tests for estimating the volume of blood found at a crime scene, and a method to analyze a suspect's DNA and link it to a victim.

Lee and other forensic investigators have helped identify bodies from mass graves in Bosnia, where political strife caused many deaths. On occasion Lee returns to his home in Taiwan to train police officers. Although retired from teaching, Lee remains active in the field of forensics. His latest project is the development of "teleforensics," a method of linking forensic examiners through telecommunications. ■

Dr. William Bass

National Science Education Standards

- Many individuals have contributed to the traditions of science. Studying some of these individuals provides further understanding of scientific inquiry, science as a human endeavor, the nature of science, and the relationships between science and society.
- Tracing the history of science can show how difficult it was for scientific innovators to break through accepted ideas of their time.

August 30, 1929–

Much of what scientists know today about corpses and human bones comes from William Bass, who created the field of forensic anthropology.

After receiving his Ph.D. in anthropology in 1961, William Bass spent 10 years teaching at the University of Kansas in Lawrence. While there, Bass excavated human remains from ancient burial grounds of Plains Indians. This led him to create one of the first knowledge bases about the life of Native Americans, one that compared life expectancies, diseases, causes of death, and lifestyles among Plains Indians.

Bass moved to the University of Tennessee in 1971 and shifted his focus to studying the remains of human bodies. He persuaded the university to let him study decomposition in a remote area of the campus. On his so-called Body Farm, Bass and his colleagues conducted research using donated bodies. They examined the effects that various conditions—buried in the ground or left in a trunk, underwater, exposed outside—would have on a body.

Criminologists and anthropologists the world over came to learn from him. His students have helped identify the remains of American soldiers in Vietnam and victims of war crimes. They have also examined the remains of the first-known humans to come to the Americas.

Now retired, Bass devotes his time to research. He also accompanies journalists, police officers, attorneys, and morticians who investigate sites where forensic anthropology could help solve a case. ■

To stimulate creative student thinking

What Killed Edgar Allan Poe?

For almost a century and a half after Edgar Allan Poe's death on October 7, 1849, people believed the writer had died an alcoholic. He had gone on drinking binges and had lost a job because of his condition. In 1996 Dr. Michael Benitez, a pathologist at the University of Maryland Medical Center in Baltimore, reviewed the case, unaware he was examining the death of a famous Baltimore resident.

Benitez received a document describing the case of "E. P., a 39-year-old man . . . brought to the emergency room in early October for evaluation of lethargy and confusion." Benitez learned that E. P. was a writer who had a history of alcoholism and possible drug abuse. He was asked to determine the cause of death.

Benitez followed standard procedure. "I looked for the lab work, the EKGs, and so forth," he says. None existed, so he guessed that the case was historical. When Benitez read medical records that attributed E.P.'s death to alcoholism, he was dubious. He'd learned from medical journals that "severe alcoholics don't go into and out of comas. They either get better or they die."

Benitez researched diseases common in the 19th century—yellow fever and malaria. Both diseases cause high fever, not a symptom of E. P. Reading the case records closely, Benitez learned E. P. experienced wide fluctuations in pulse, heavy perspiration, delirium, coma, and confusion. He suspected the common disease of rabies, a fatal viral infection of the brain and nervous system caused by the bite or saliva of a rabid animal. E. P.'s symptoms matched those of rabies. In addition, says Benitez, "the clinical course of rabies is four days, which is how long it took [this person] to die." More convincing evidence came to light: E. P. had asked for water, but couldn't swallow it. Rabies is also called hydrophobia, a Greek word meaning "fear of water." The inability to swallow liquids is one of the disease's final symptoms.

Benitez's deduction that Edgar Allan Poe had died of rabies shocked the literary world. Yet it made sense. Poe's friends wrote that he had given up drinking. And Poe loved animals and kept cats as pets. Although there is no record of Poe receiving an animal bite, Benitez noted that only 27 percent of recent rabies victims ever remember the bite—and that the disease can remain in the nervous system for up to a year before symptoms surface.

Did rabies kill Edgar Allan Poe? Dr. Benitez believes so and published his findings in the *Maryland Medical Journal.* His careful detective work indicates that the writer died of rabies, a fearsome and sometimes fatal disease caused by the bite of an animal. ■

Professional Resources

Peer groups, programs, associations, and more

National Science Education Standards

National Research Council

National Academy Press

The *National Science Education Standards,* developed by thousands of teachers, scientists, science educators, and other experts, offers a comprehensive guide to teaching science in schools today. The standards and teaching practices in this resource provide a coherent vision of what it means to be scientifically literate, describing what all students in grades K–12 should understand and be able to do in various science categories. Order from

National Academy Press
2101 Constitution Avenue, NW, Lockbox 285
Washington, DC 20005
(800) 624-6242 or (202) 334-3313
www.nas.edu ■

Educational Resources

National Science Teachers Association

1840 Wilson Blvd., Arlington, VA 22201-3000
(703) 243-7100, www.nsta.org

The NSTA publishes newsletters, five journals, many books, a children's magazine called *Dragonfly,* and many other publications. NSTA conducts national and regional conventions in the spring and fall. Their Web site links to many additional resources.

Also be sure to check out the Science Teacher Association in your state. State conferences provide quality workshops and teaching materials, as well as exhibits from publishers and local organizations.

Kathy Schrock's Guide for Educators

http://school.discovery.com/schrockguide/

At this site, sponsored by Discovery Channel, you will find a categorized list of sites useful for both curriculum and professional development. The site is updated regularly for teaching a wide range of topics.

American Academy of Forensic Scientists

www.aafs.org

This professional society is dedicated to the application of science to the law. Membership includes physicians, criminologists, toxicologists, attorneys, dentists, physical anthropologists, document examiners, engineers, psychiatrists, educators, and others who practice and perform research in the many diverse fields related to forensic science.

Elsevier Science

655 Avenue of the Americas, New York, NY 10010-5107
(212) 633-3730, www.elsevier.nl/homepage

This leading publisher of scientific books and journals offers numerous publications in the field of forensic science and medicine.

National Institute for Science Education

1025 West Johnson Street, Madison, WI 53706
(608) 263-9250, fax: (608) 262-7428, www.wcer.wisc.edu/nise

Funded by the National Science Foundation, NISE is a project of the Wisconsin Center for Education Research. Founded in 1995, NISE's goal is ensuring that all students leave the educational system with an ability to make informed decisions about science-related matters they encounter in daily life. Using Web sites, publications, and workshops, NISE strives to develop active contextual learning that builds a strong science foundation and encourages young people to become lifelong learners.

Local Police Department

Representatives from local police departments may be available for speaking appointments in schools. Most precincts also have information about local programs for safety, crime prevention, and fingerprinting. Check with your area police or sheriff's department to find out about opportunities you can share with your students. ■

Catalog

Flynn Scientific

www.flinnsci.com
P.O. Box 219, Batavia, IL 60510
(800) 452-1261

Questions and Answers

Common questions about Forensics from curious kids

Discovery Channel School

Glad You Asked

Students will love knowing the answers.

What is forensics?

Forensic science is applying scientific knowledge to legal matters. Within the field of forensics there are many specialties. For example,

- forensic anthropologists analyze bones.
- forensic biologists work with blood, saliva, DNA, and other biological substances.
- forensic chemists analyze ink, paint, fibers, and bloodstains.
- forensic entomologists study insects found on dead bodies and at crime scenes to learn the time of death and possibly if a body has been moved.
- forensic pathologists study the causes and results of diseases and injuries.

If a police department does not have access to forensic scientists, they send evidence to the FBI crime lab for analysis.

What is DNA profiling?

DNA profiling is the attempt to link cells found at the scene of the crime with a suspect. A DNA profile looks like a bar code and identifies coded information about the genetic makeup of the person. DNA, or deoxyribonucleic acid, is part of the chromosomes, the long threadlike molecules in the nucleus of every living cell. DNA consists of nitrogenous bases arranged in long chains that carry the genetic code, information that determines every trait from hair color to susceptibility to certain diseases. Every person's sequence of the code is unique, except for identical twins, who have identical DNA. Because DNA has unique characteristics, forensic scientists can use it to make a genetic profile.

What is the chain of custody?

A court of law must be assured that no one has tampered with any crime-scene evidence. So all evidence must be carefully documented. The chain of custody details who has had access to evidence between the time it was first collected and the time it is presented in court. Without detailed accounting, evidence may be considered inadmissible, or unusable in the case.

Are fingerprints always visible?

Latent fingerprints, such as those found on a note or glass, are invisible or very faint. Such prints cannot be seen without dusting, a process that makes them visible. Investigators may use the technique of fuming fingerprints. In this process, they place the print and superglue in a sealed container. The chemical cyanoacrylate, released in the superglue fumes, will stick to the oils in the fingerprint. Within 48 hours the print appears white.

How do experts match fingerprints?

Fingerprints are composed of straight and curved lines. To match prints, investigators look for points of similarity, such as ridge endings, dots, short ridges, and bifurcations and trifurcations, or points where the lines split in twos and threes. Most fingerprint experts accept eight points of similarity to make a match. If investigators can't match a print, they may send its scanned image to the FBI lab for identification. Some police departments use the Automated Fingerprint Identification System (AFIS) to scan a computer database and identify prints. ■

Quiz Center

Extend questions and answers into a classroom activity. Create quizzes and worksheets based on any of the above topics with the **Quiz Center** tool at **discoveryschool.com.** ■

Name ______________________ **Class** ______________________

Finding Information in Fiber and Hair Samples

You can tell a fiber from a hair. Observe the scales on the outside of the hair strand, the roots that attach it to the scalp, and the medulla, or center part of the strand. Synthetic fibers don't have these features. Under the microscope a fiber appears solid or intertwined with many other strands of thread.

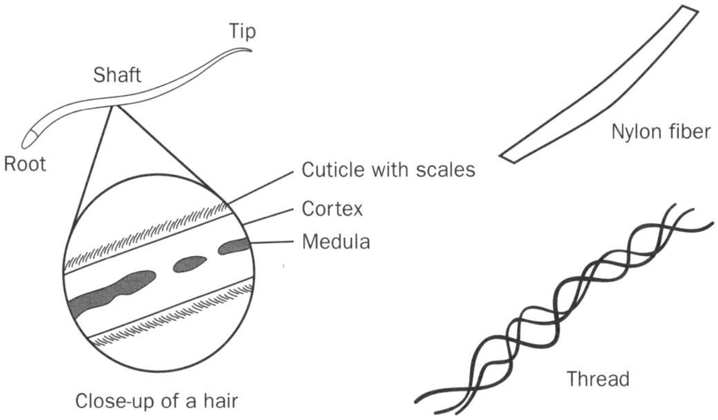

Close-up of a hair

Look at these drawings of microscopic images. Are they animal fur or human hair? Notice the differences between roots, scales, and medulla.

Forensic scientists can tell how a hair was removed. Notice the root on the hair pulled from the scalp. When the hair root has tissue attached, investigators can extract DNA and follow this trail of evidence.

Roots

Mature (telogenic) human root (club shaft)

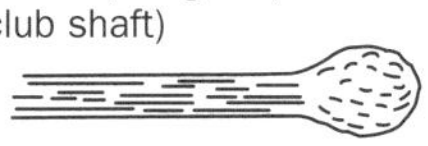

Cat root (often frayed at base)

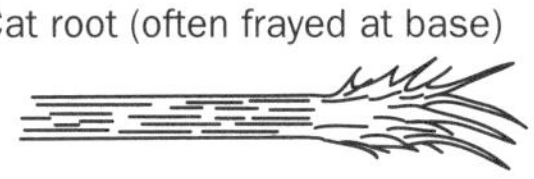

Dog root (often spade-shaped)

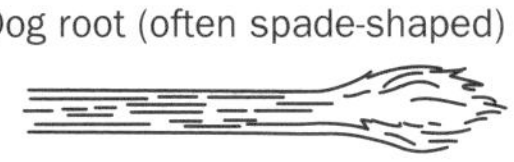

Medulla Types

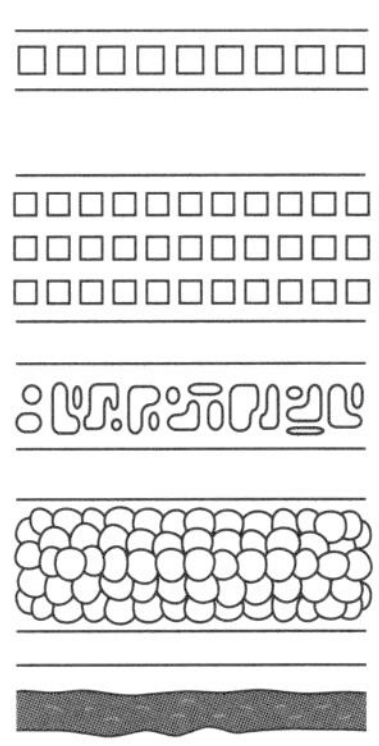

Uniserial (rabbit)

Multiserial (rabbit)

Vacuolated (dog, fox, common)

Lattice (deer)

Amorphous (human, common)

Scale Patterns

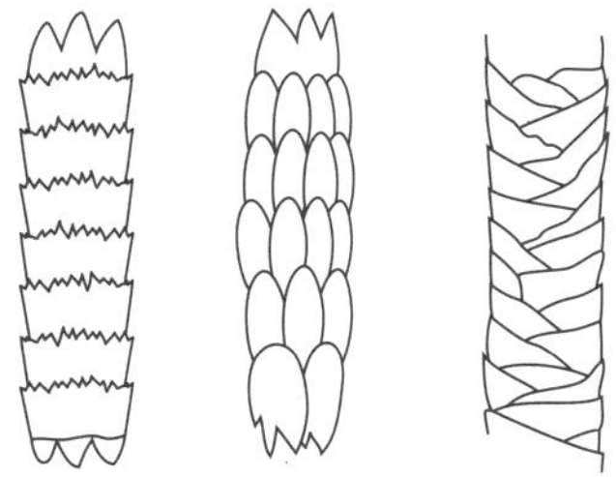

Coronal (mouse) — Spinous (cat) — Imbricate (human)

Tales of Hair

An examiner can tell if hair was dyed or bleached by looking at the cuticle, or outside layer of the hair. Dyed hair shows a color residue. Bleached hair looks yellowish because the hair pigment has been removed.

Anagenic human root with follicular tag (forcibly removed)

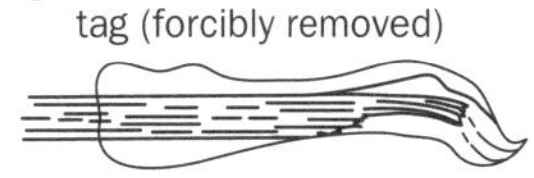

Hair tells other tales, too. Investigators found arsenic in the French emperor Napoleon Bonaparte's hair more than 130 years after he died. The discovery confirmed the theory that Napoleon was poisoned to death.

Collect a strand of your own hair from a brush or comb. After examining it under a microscope, make a diagram in the space below. What does the cuticle look like? Does the medulla have a distinctive pattern? What shape is the root? ■

Name ______________________ Class ______________________

Searching for Fingerprints

A good investigator knows where to look for fingerprints, which can help place a suspect at the scene of a crime. In one case, bank robbers used an apartment to plan their robbery. They thought they had wiped away all their prints, but they forgot to run the dishwasher. The police searched for clues to show that the robbers had met together to plan the crime. When investigators found prints on the dirty dishes, the evidence supported the fact that they were together in one place.

Fingerprints come in three large groups of patterns: arch, loop, and whorl. The three major groups are divided into smaller groups, based on smaller variations. Study the diagrams below and see if you can identify types of fingerprint patterns.

What kinds of fingerprint patterns do you have? Make your own prints. Roll each finger over a black stamp pad and then press each one into the labeled diagram below. Make careful observations of your prints: Look for loops, whorls, and arches. Can you identify your fingerprint patterns by comparing them with the samples on this page? Compare other students' fingerprints. Do you and your classmates have any patterns in common? ■

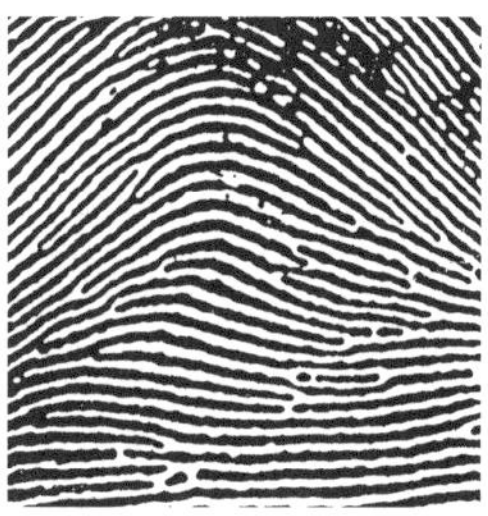

Left pinky	ring finger	middle finger	index	thumb
Right pinky	ring finger	middle finger	index	thumb

Science Projects

Encourage students to discover their own answers

National Science Education Standards

- Students in grades 5–8 should be helped to acquire scientific knowledge, use high-level reasoning, apply their existing understanding of scientific ideas, and communicate scientific information.

General Science Project Guidelines

Science projects should provide students with an opportunity to apply what they have learned about the scientific process and inquiry. Parents and teachers should serve as resources, providing guidance and reaction, but students must own the project, making their own decisions regarding the question, data collection, variables they will control, depth of study, and the format of the final report or display.

When designing their project or investigation, students should:

- Use appropriate tools and techniques to gather, analyze, and interpret data.
- Develop descriptions, explanations, predictions, and models using evidence.
- Think critically and logically to identify the relationships between evidence and explanations.
- Recognize and analyze alternative explanations and predictions.
- Communicate procedures and explanations.
- Use mathematics.

Science Project Starter Questions

Research

- How might RNA, rather than DNA, be used to identify an individual? How about a person's proteins?
- How did people in ancient civilizations solve crimes? What techniques of forensic science did they use?
- What simple chemical tests can be used to determine the presence of sugar or proteins at the scene of a crime?
- How does your local police department research fingerprints?

Historical Research

- What kinds of evidence were used in American and British court cases in the early 1900s?

Modeling

- What types of fingerprints do your classmates have? Create groups of types within the class.
- Can you develop a new, improved system of fingerprint classification?
- How can footprints be used to identify different individuals?
- What systems could be developed to determine a person's height and weight from length of arm, leg, hand, head, or other similar body measurements?
- What is required to create a database of physical characteristics to identify your classmates? ■

Young Scientist Challenge 2001

Discovery Channel School provides assistance to teachers and students at http://school.discovery.com/sciencefaircentral with an online science fair handbook. Remember that the objective of a science project is to learn more about science, not to produce an elaborate display or win a prize.

For students who want the ultimate challenge, enter one of these projects or any others in the Discovery Young Scientist Challenge. Visit http://school.discovery.com/sciencefaircentral/dysc/index.html to find out more. ■

Testing

How to assess student performance and concept mastery

Performance-based evaluation enables the teacher to assess student understanding of concepts through the use of activities, experiments, projects, and journal writing.

Performance assessment consists of a task and rubrics—a scoring method used to judge the quality of student performance.

Student Journals

Activities and Experiments in *Forensics A-Z* can be used as both teaching and testing experiences. Student journals enable you to use these experiences to obtain both pre- and post-evaluation of student learning. Prior to each experience, pose a question and ask students to write their thoughts about that question. During the experience, have them record notes and data. Following the experience, students reflect on what they did and what they now know about the question asked. This pre- and post-evaluation will provide a documentation of student growth in understanding the big ideas about forensics.

Developing a Scoring Rubric

Because student output, such as a multimedia slide show, is more complex than a written assignment, assessing a student's work may seem more difficult. In a project such as this, students must be able to express their learning clearly, demonstrating not just what they know but also their ability to communicate it to others. An important part of assessing the work consists of students evaluating themselves and each other in addition to evaluation by the instructor. The tool that makes it possible to evaluate work in this student-centered context is a rubric.

A rubric is simply a list of criteria by which student output is judged. It defines the attributes of a quality project, provides the criteria by which a project will be assessed, and gives specific indicators that describe various levels of achievement for each of those criteria. In this respect, a rubric is a concrete guide or checklist that students can refer to as they work, besides being a measuring device for the teacher.

For example, a generic rubric for a complex project might look something like this:

	Needs Work (1 point)	Average (2 points)	Best (3 points)
Content	Data is inaccurate in places or key pieces of information are missing.	Data is complete, but inaccurate in places.	Data is both accurate and comprehensive.
Presentation	Is drab and ordinary.	Is entertaining and interesting.	Shows creativity and is entertaining; shows that the student had fun putting it together.
Effort	Project exhibits minimal effort to fulfill the assignment.	Project shows evidence of personal interest in the topic and thoughtful analysis.	Project shows evidence of personal interest, thoughtful analysis, and creativity in the information's presentation.
Publishing	Text has numerous errors in grammar, usage, spelling, or punctuation.	Text has few errors in grammar, usage, spelling, or punctuation.	Text has no errors in grammar, usage, spelling, or punctuation and utilizes higher level vocabulary and complex sentence structure.

This example is just a suggestion. Keep in mind that you know your students best and you know their academic needs. As the teacher, you will need to develop your own rubric for your class so that it reflects:

- the age and prerequisite skills of your students
- the specific goals you have for your class
- the complexity of the content topic selected
- the level of achievement you expect

The science project questions on page 23 can serve as long-term tasks or the basis for multimedia or similarly complex projects.

Forensics CD-ROM

Have your students create their own multimedia projects on forensics or explore movie clips on the *Forensics* CD-ROM by Discovery Channel School (network version #760520; single user #760967). To order call (888) 892-3484. ■

Unsolved Mysteries

Science awaits solutions

Who Was the Ice Man?

Will your students be the ones to find the answers someday?

About 5,300 years ago a hunter struggled across the frozen slopes in what are now the Italian Alps. In the blinding snow and piercing winds the man stumbled into a glacier and froze to death. During the relatively warm winter of 1991, this glacier melted enough to reveal the hunter's body. It was almost totally intact. Alongside the man were a copper ax, a bow, and flint-tipped arrows. For scientists, the discovery of the "Ice Man" opened a window into the Stone Age.

Scientists kept the hunter in a refrigerated viewing chamber at the South Tyrol Museum of Archeology in Bolzano, Italy. While scientists studied his tools, clothing, and other artifacts, the Ice Man remained frozen, his life and death as much a mystery as when he was discovered . . . until late September 2000, that is.

On the morning of September 25, a team of forensic anthropologists slowly raised the temperature in the display case. After 12 hours, the body had thawed and the team, dressed in surgical scrubs, began their work.

The scientists scraped off bone enamel, chipped away bone samples, and snaked a scope through the man's digestive system. They took the samples to more than a dozen research institutes around the world.

What will the anthropologists look for in the samples? A forensic expert in Scotland will look at bone and blood samples to determine the cause of death. Was it accidental or natural?

In Switzerland scientists will analyze chemical deposits on the man's teeth to learn more about his environment. What was the atmosphere and water like 5,000 years ago?

Scientists in Italy and Great Britain will examine the microbes in the intestines. They hope to find clues to the diet of Stone Age people. Did they eat mostly plants or animals?

Elsewhere, scientists will look at his DNA, which could link the Ice Man to people living in the Alpine region today. Their analysis of his DNA may focus on a specific family tree.

Finally, scientists will try to determine if tattoos on the Ice Man's ankles and calves were a prehistoric method of acupuncture.

Initial results of the research will be released in 2001, but scientists expect their work to continue for many years. Meanwhile, the Ice Man has resumed his long frozen sleep in a viewing room. Little did he know on that day he fell into a glacier that someday he would have an important place in the world. ■

V Vocabulary

Key Forensics words and terms

Discovery Channel School

Use for your own reference or copy and distribute to students.

autopsy—the examination of a body to determine time and cause of death

ballistics—the science of firearm projectiles

blood types—the main categories of blood found in human beings. There are four major blood groups—A, B, O, and AB. These are further divided into Rh Positive and Rh Negative types: A (Rh Positive); A (Rh Negative); O (Rh Positive); O (Rh Negative); B (Rh Positive); B (Rh Negative); AB (Rh Positive); and AB (Rh Negative).

chromosomes—thread-like strands of DNA and protein material found in the nucleus of living things that carry hereditary information

criminologist—a forensic scientist who analyzes and interprets physical evidence to link suspects to crimes

DNA—deoxyribonucleic acid, the substance of which chromosomes are made

evidence—something that is helpful in forming a conclusion; often it supports witness testimony in a court of law

forensic—relating to or used in courts of law or public argumentation

forensic anthropology—a science that applies the study of the structure, evolution, and culture of human beings to criminal and legal cases

forensic medicine—a science that applies medical knowledge to criminal and legal cases

forensic science—a discipline that examines and evaluates evidence in criminal and legal cases

galvanic skin response—a measure of electrical conductance of the skin that is the basis for the measurements taken by the polygraph or lie detector

gene—a hereditary unit passed from parent to child that determines a physical characteristic

latent—something invisible to the naked eye, such as a fingerprint or shoeprint

odontologist—a scientist who studies teeth, bite marks, and dentures. A forensic odontologist applies this knowledge to criminal cases.

pathologist—a medical doctor who studies the causes and results of diseases and injuries that lead to death

physical evidence—all material gathered at a crime scene

polygraph—an instrument that measures changes in a person's heartbeat, blood pressure, breathing, and perspiration while being questioned

questioned-document examiner—a specialist who studies handwriting to determine forgeries and other questions of legality

reagent—a substance used to produce a chemical reaction that can detect, measure, examine, or produce other substances

rifling—grooves cut into the inside of a gun barrel that cause a bullet to spin and that produce marks on the bullet that can be traced back to an individual weapon

striations—scratch marks on the surface of a bullet

toxicology—the study of poisons and their effects ■

Writing Ideas

Reproduce and distribute for instant Forensics-related writing assignments

Assess students' understanding of forensics through their writing on the topic.

Clue Us In

You're a crime-scene expert and you've been called to the scene. You spring into action and collect little pieces of evidence. Your trained eye takes in everything. Once you're back at the office, you write a report of what you've found and what you believe each specimen teaches. At right is a list of items you might find at a crime scene. Choose at least four, but no more than seven, from the list. Re-create the crime and explain how forensic tests may help crack the case. ■

- lipstick tube
- cigarette butt
- empty wallet
- handwritten note
- an empty safe
- apple core
- blowflies
- footprint
- bullet hole in wall
- matchbook
- carpet fibers
- synthetic hair
- human hair
- bullet cartridge
- 3-foot-long piece of rope
- wrecking bar

On the Beat

You're a police reporter for your local newspaper. You hear about a fire at a local donut shop that police suspect may be arson. Write a five-paragraph article with an attention-grabbing headline for the morning newspaper. How many different experts are involved in investigating the fire? What evidence did police find leading them to think this fire was arson? Do some research on causes of fires and arson to be sure your facts are solid. ■

The Defense Never Rests

You're a criminal defense attorney who has just been hired by someone about to go on trial. In court you'll have to cross-examine forensic specialists about the evidence gathered at the scene. (See "Clue Us In" above.) As you go through the items, write two or three questions that you can ask about the accuracy of the forensic tests. ■

Create Your Own Sherlock

Sherlock Holmes is the main character in Sir Arthur Conan Doyle's classic fictional series of mysteries. This British detective is always on the trail of a mystery that has other investigators baffled, and with the help of his assistant, Watson, he always cracks the case.

Dream up your own lead character—it could even be you—who uses forensic evidence to solve a case. Write a mystery in which you explain who the investigator is and how he or she stumbles onto the case and ultimately solves it. Be sure to describe what evidence the investigator finds, any other characters that may help this hero come to a conclusion, and how the lead character cracks the case. ■

Do-It-Yourself Detective

You notice that a page is missing from your favorite book! How can you go about finding out who did it? Write a step-by-step account of what you would do to catch the culprit, supposing you have any kind of tool at your disposal. ■

Forensic Anthropologists: At Work Around the World

Forensic anthropologists are detectives who try to solve cases involving human remains.

Forensic anthropologists use the same techniques to learn about modern victims as they use to study early humans (see **Unsolved Mysteries,** page 25). From bones, a forensic anthropologist may discover much about a person.

- **Gender:** The pelvis, base of the skull, the forehead, and the jaw differ in size and structure in males and females.
- **Age:** Examine specific bones, joints, and teeth. Different bones fuse together when a person reaches a specific age. For example, by age 14 the bones in your elbow should connect.
- **Height:** based on the lengths of leg and arm bones. If you find certain bones intact, you can use calculations to find information about the person it once belonged to. If you have the femur, or thigh bone, of a Caucasian male (to be determined in other tests), the calculation is: length of femur × 2.38 + 61.41 cm = height.
- **Disability:** by the wear on the bones at certain points. A person who walks with a limp may have leg bones that are worn away in some sections.
- **Race:** by examining the width and height of the nose and other facial features. The bones around the nose are wider on non-European individuals.
- **Injuries:** if bones are fractured or damaged.
- **Cause of death:** by evidence such as stab marks, marks on the skull, broken bones, or bullets in or near a body.

The map on these pages features sites at which forensic anthropologists have examined and helped identify human remains. Their findings have helped solve crimes, as well as historical mysteries that help us better understand the past.

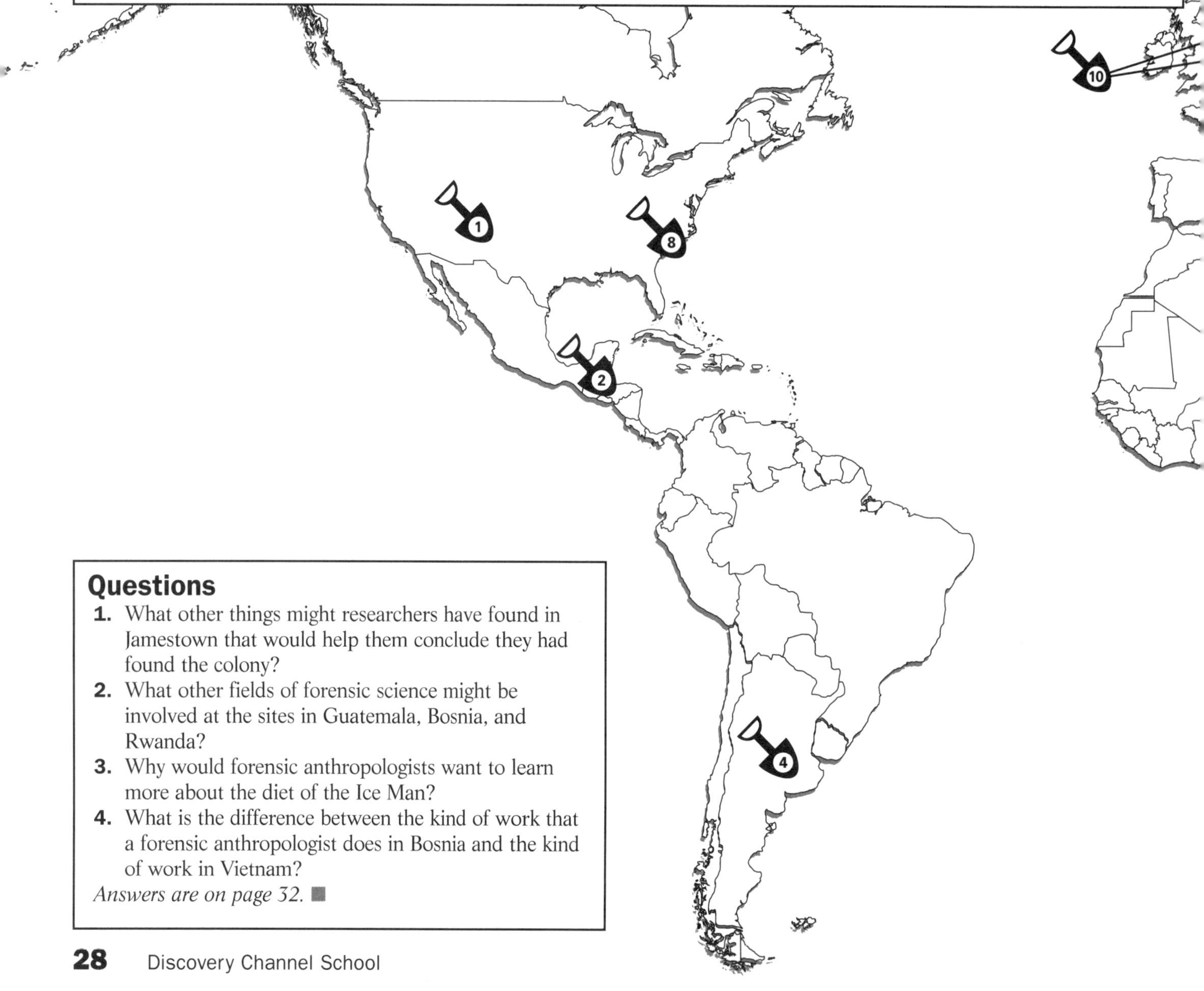

Questions

1. What other things might researchers have found in Jamestown that would help them conclude they had found the colony?
2. What other fields of forensic science might be involved at the sites in Guatemala, Bosnia, and Rwanda?
3. Why would forensic anthropologists want to learn more about the diet of the Ice Man?
4. What is the difference between the kind of work that a forensic anthropologist does in Bosnia and the kind of work in Vietnam?

Answers are on page 32.

Discovery Channel School

X Marks the Spot

X

1 Four Corners area, U.S.—The ancient Anasazi Indians disappeared more than a thousand years ago. Some evidence points to cannibalism among warring tribes.

2 Guatemala—Mass graves of victims of a 40-year civil war

3 Italy—Bone and skin samples from the "Ice Man" (see **Unsolved Mysteries,** page 25), discovered frozen in a glacier

4 Argentina—The remains of hundreds of people who disappeared during political upheaval in the 1970s and 1980s

5 Bosnia—Mass graves of victims of a civil war in the 1990s

6 Rwanda—Mass graves of victims of a civil war in the 1990s

7 Vietnam—The remains of American servicemen declared missing in action during the Vietnam War have been found.

8 Virginia—Skeletons and burnt pieces of wood date back to the late 1500s or early 1600s, leading researchers to believe that they belong to residents of Jamestown, the first British colony in North America.

9 Russia—Skeletal remains of the Romanovs, Russia's last royal family, are uncovered 61 years after they were killed in 1918.

10 Denmark and England—Well-preserved bodies buried in bogs, or wetland areas, are victims of ritual sacrifices that took place more than 1,500 years ago.

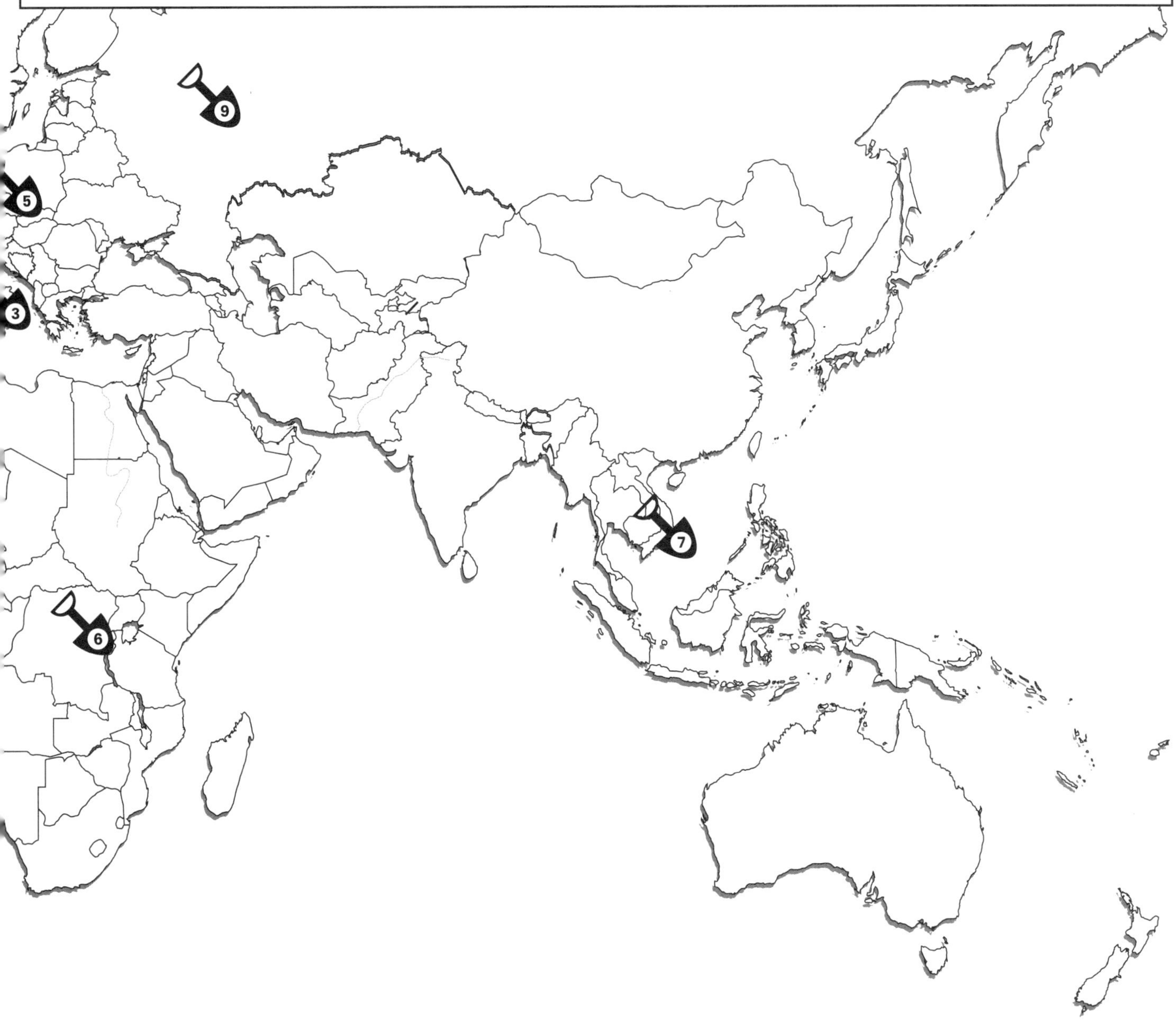

Y Year After Year

Significant science, traced over time

Here are some key dates in the history of forensic science. ■

5th century BC—Prehistoric humans leave fingerprints on paintings and rock carvings.

8th century BC—Chinese use fingerprints to catalog the identity of authors and artists.

AD 100—The Roman attorney Quintilian uses bloody palm prints from a crime scene to prove the innocence of a wrongly accused suspect.

1686—Marcello Malpighi, professor of anatomy at the University of Bologna, Italy, describes the different characteristics of fingerprints.

1784—The first recorded use of matching evidence: John Toms of Lancaster, England, is convicted of murder, based on torn newspaper pages found in his pistol and his pocket.

1835—Henry Goddard, of Scotland Yard, is the first detective to use bullet comparison to solve a case. He traces a flaw in the bullet's shape to a gun owned by a suspect.

1877—Scientist Thomas Taylor publishes an article in *Scientific American* suggesting that fingerprints could be used for identifying criminals.

1880—The first recorded use of fingerprints to solve a crime: Henry Faulds, a physician in Tokyo, Japan, uses fingerprints to clear an innocent suspect and identify the burglar.

1883 —French criminologist Alphonse Bertillon is the first to use physical measurements and photographs to identify criminals. His system includes front- and side-view photos, as well as the subject's height, length of an index finger, an arm, and one foot.

1889—Alexandre Lacassagne, professor at the University of Lyons, France, is the first to match individual bullets to specific gun barrels.

1891—Hans Gross, professor of criminal law at the University of Graz, Austria, publishes *Criminal Investigation,* the first book to describe the uses of physical evidence for solving crimes.

1896—Sir Edward Henry publishes *Classification and Uses of Fingerprints,* the first fingerprint classification system. Henry is appointed head of Scotland Yard in 1901.

1901—First official use of fingerprints in the United States by the New York City Civil Service Commission.

1905—President Theodore Roosevelt establishes the United States Federal Bureau of Investigation (FBI).

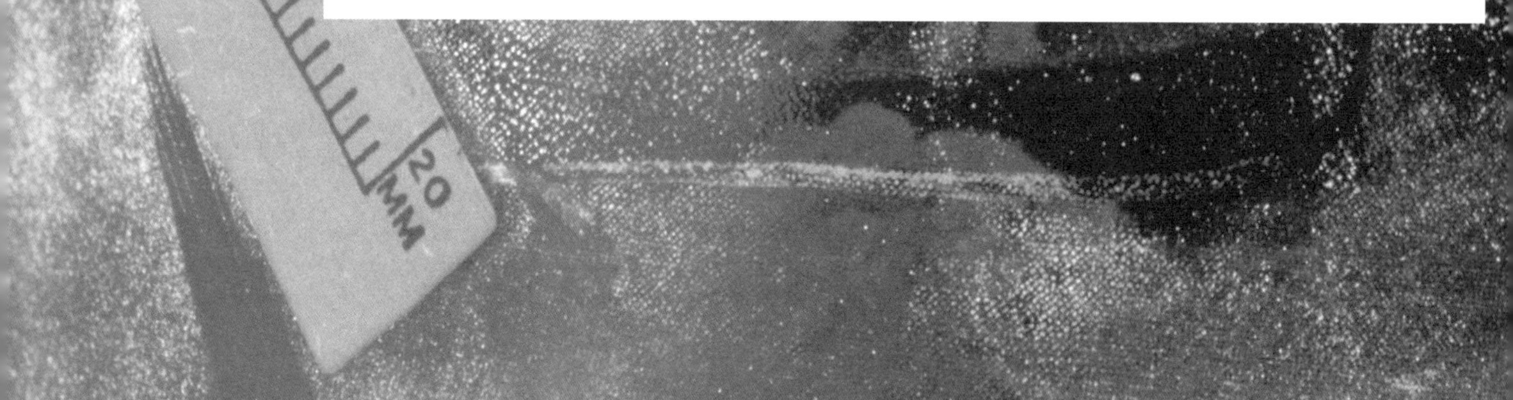

Year After Year

1910—The first police crime laboratory is established at the University of Lyons, France.

1916—Detective Albert Schneider of Berkeley, California, first uses a vacuum tool to collect crime-scene evidence.

1921—American scientists John Larson and Leonard Keeler design and build the first polygraph, or lie detector.

1923—The U.S. Supreme Court rules that polygraph tests are inadmissible in trials.

1924—The first U.S. crime laboratory is established in Los Angeles, California.

1930—The FBI sets up a national fingerprint file in the U.S.

1950—The American Academy of Forensic Science (AAFS) is founded in Chicago, Illinois.

1974—The technique for detecting gunshot residue through electron microscopy is developed.

1977—The FBI introduces the Automated Fingerprint Identification System of computerized scans of fingerprints.

1984—Sir Alec Jeffreys develops the first DNA profiling test at University of Leicester, England.

1986—In the first use of DNA to solve a crime, Jeffreys uses DNA to identify the bodies of two females murdered in England. DNA profiling also establishes the innocence of the original suspect.

1987—Investigators' first success with DNA testing leads to the conviction of Tommy Lee Andrews, a violent criminal in Orlando, Florida.

1996—The FBI introduces computerized searches of the Automated Fingerprint Identification System (AFIS) fingerprint database. Scanning devices allow interdepartmental submissions of fingerprints.

1998—An FBI DNA database is in use, enabling interstate cooperation in linking crimes.

1999—The FBI upgrades its computerized fingerprint database and implements the Integrated Automated Fingerprint Identification System (IAFIS), allowing paperless submission, storage, and search capabilities directly to the national print database maintained at FBI headquarters. ■

Z

Zingers

Thought-provoking tidbits

Discovery Channel School

The Fascinating World of Forensics

A great way to begin or end class.

Murders at Mealtime

An unusual crime-solving organization meets regularly for lunch in Philadelphia, Pennsylvania. The Vidocq Society is named in honor of Eugène François Vidocq, the founder of the first detective force. Vidocq Society members work to solve unsolved crimes. Most members are retired detectives, attorneys, and other professionals who volunteer their time, forensic experience, and talents.

Cracking Creature Crimes

At the National Fish and Wildlife Forensics Laboratory in Ashland, Oregon, crime has a different face—or paws, claws, or feathers. Founded in 1989, the world's first wildlife crime lab examines cases of illegal trade in wildlife or animal parts, such as illegal or fake caviar, and tracing shawls made from Tibetan antelope hide. The lab must identify an item and connect it to a crime scene, a victim, and a suspect. It's not easy. "In human forensics labs, they're dealing with one species," says director Gary Espinoza. "Ninety percent of what we do is determining what species a thing is."

The Telltale Toothpick

Police in Boston, Massachusetts, solved a homicide case even though the only evidence available was a toothpick. Using scientific methods pioneered by Dr. Henry Chang-Yu Lee (see **Noteworthy People,** page 17), they tested the DNA from cells found on the toothpick and linked it to a suspect. ■

Credits

Writers: Scott Ingram, Rhonda Spidell
Editors: Justine Ciovacco, Monique Peterson
Illustrations: Thomas Gagliano
Cover: ©TEK Image/ Science Photo Library

Answers

Who Ya Gonna Call?, page 9
1. h **2.** b **3.** f **4.** a **5.** c **6.** d **7.** g **8.** e

R U Clueless? page 9

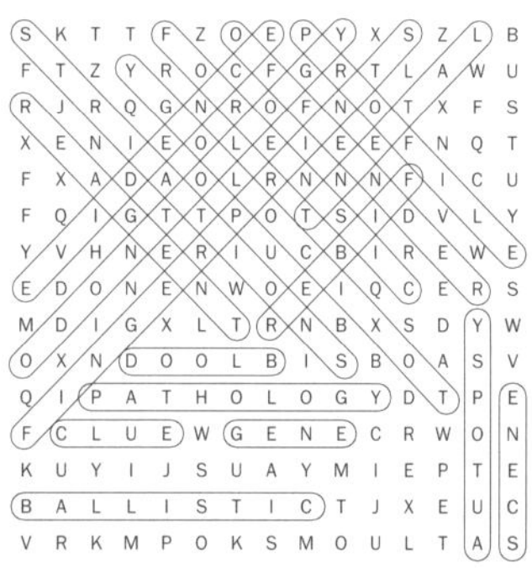

Homework Helpers, page 10
The life cycle of the blowfly is approximately 20 days. Only the pupal casings were left, so we know that the deer was dead for more than 19 days.

The life cycle of the cheese skipper is approximately 40 days. Adult cheese skippers were found on the carcass, so we can assume that the deer was dead at least 40 days. We can also figure that the daily temperature was mild to warm so the life cycles would not have been delayed.
No evidence of the lesser housefly was present, so we can assume that the carcass hasn't been decomposing for as long as 48 days.

The deer has been dead for at least 39 days, but less than 50, which would prove that it was shot before hunting season opened.

Crime Rhyme, page 13
1. claws (or paws) laws **2.** shot spot **3.** right bite
4. print hint **5.** drugs and bugs **6.** gene dean **7.** fab lab
8. shoe clue **9.** flame blame **10.** lie guy **11.** choose clues
12. top cop

X Marks the Spot, pages 28–29

1. They found the bodies of townspeople and matched them—using DNA analysis and studying their bones—with records of people who lived in the colony. (Accept other reasonable answers.)
2. The fields of forensic anthropology, ballistics, entomology, odontology, DNA fingerprinting. (Accept other reasonable answers.)
3. In learning what the Ice Man ate, forensic anthropologists could find out what foods grew wild or were raised in the region 5,000 years ago. This would provide insight into the climate, possible nutritional deficiencies that contributed to health problems, and other prehistoric elements of life.
4. The bodies found in Bosnia have been dead for a shorter period than those found in Vietnam. Finding the cause of death for the bodies in Bosnia is important for investigation into war crimes. Anthropologists there are examining the remains of citizens of their own country. The main task for those studying the bodies from Vietnam is identification of skeletal remains. American army officials and civilians are trying to determine if the bodies are the remains of American or Vietnamese servicemen.